YOUR KNOWLEDGE HAS VALUE

- We will publish your bachelor's and master's thesis, essays and papers

- Your own eBook and book - sold worldwide in all relevant shops

- Earn money with each sale

Upload your text at www.GRIN.com and publish for free

Bibliographic information published by the German National Library:

The German National Library lists this publication in the National Bibliography; detailed bibliographic data are available on the Internet at http://dnb.dnb.de .

Imprint:

Copyright © 2017 GRIN Verlag, Open Publishing GmbH
Print and binding: Books on Demand GmbH, Norderstedt Germany
ISBN: 9783668542693

This book at GRIN:

http://www.grin.com/en/e-book/377109/a-study-on-th-effect-of-organic-inhibitors-on-the-corrosion-of-steel-and

Narayana Hebbar

A Study on th Effect of Organic Inhibitors on the Corrosion of Steel and Zinc

GRIN Publishing

A Study on the Effect of Organic Inhibitors on the Corrosion of Steel and Zinc

Viva-voce examination of Ph.D. Degree in chemistry of Visveraya Technological University, Belagavi

Presented by
Narayana N Hebbar
Research Scholar
Srinivas School of Engineering, Mukka

Contents

Introduction

- Mild steel and zinc are most widely used engineering materials especially for structural   applications.

- They are exposed to various aggressive environments during their usage and hence undergo corrosion.

- Corrosion of steel and zinc materials and their protection have become fundamental, academic and industrial concern

- The corrosion of metals received a prime attention of scientists since it has many serious consequences including economic, health, safety, technological and cultural consequences all over the world.

- If economic loss alone is considered, the costs associated with corrosion can be estimated with reasonable accuracy. The typical estimations were US $275 billion dollars in1998 and US$364 billion dollars in 2005, draining about 3.1% of the GDP from the US economy [1-6].

- These numbers alone justify much greater need for development of effective corrosion protection techniques

- Various methods have been employed in controlling their corrosion. The use of chemical inhibitors is one of the best practical methods in controlling the corrosion.

- Various types of chemical inhibitors like organic and inorganic compounds, pharmaceuticals drugs, dyes and even plant extracts are developed to control corrosion.

- But the practice of corrosion inhibition is greatly influenced by new regulations that have been developed to check the use of toxic compounds and environmental damage resulting from the use of such compounds.

- Hence there is a trend to replace some widely used inhibitors such as chromates and cyanides because of their toxicity, environmental damage and pollution caused by these chemicals.

- Although many organic inhibitors were developed as corrosion inhibitions for mild steel and zinc, a little work has been done in detecting and using electro active compounds such as dyes, drugs and natural extracts as corrosion inhibitors for steel and zinc.

- The aim of the present proposal is to find the new class of non toxic inhibitors for controlling the corrosion of steel and zinc in acidic medium and establish the mechanism of corrosion by analytical and theoretical methods.

Literature Survey

- Mild steel has been extensively used under different conditions in chemical and allied industries in handling alkaline, acid and salt solutions. Chloride, sulphate and nitrate ions in aqueous media are particularly aggressive and accelerate corrosion. One way of protecting mild steel from corrosion is to use corrosion inhibitors. Most of the well-known inhibitors are organic compounds containing nitrogen, sulphur and/or oxygen atoms. It has been observed that most of the organic inhibitors act by adsorption on the metal surface [8 - 12].

- Zinc is a metal with numerous industrial applications and is mainly used for the corrosion protection of steel. The dissolution behavior of zinc in acidic and nearly neutral media is known to be inhibited by nitrogen and sulfur-containing organic compounds. Such compounds contain electron-donating groups that decrease the corrosion rate by increasing the hydrogen over voltage on the corroding metal. Studies of the effect of organic additives on the corrosion rate of zinc have been the subject of many investigators. The efficiency of aldehydes and amino acids as inhibitors of corrosion for different metals in different corrosive environments has been studied by several workers, and their inhibition efficiency is due to the formation of a protective layer or film on the metallic surface. The role of this film is to isolate the corroding metal from the corrosive medium [13-25].

- Many works have been done for controlling the corrosion of steel and zinc metals in acidic and neutral medium some of them are discussed below

- The corrosion inhibition of mild steel in 0.5 M hydrochloric acid solutions by some new hydrazine carbodithioic acid derivatives namely N0-furan-2-yl-methylene-hydrazine carbodithioic acid (A), N0-(4-dimethylamino-benzylidene)-hydrazinecarbodithioic acid (B) and N0-(3-nitro-benzylidene)-hydrazine carbodithioic (C) was studied. The inhibition efficiency obtained by these compounds is increased by increasing their concentration. The inhibition efficiency follow the order C > B > A. Polarization studies show that these compounds act as mixed type inhibitors in 0.5 M HCl solutions. [26].

- The three benzimidazole derivatives, namely, 4-(phenyl)-5-[(2-methyl-1*H*-benzimidazol-1-yl)methyl]-4*H*-1,2,4-triazole-3-thiol, 4-(4-methylphenyl)-5-[(2-methyl-1*H*-benzimidazol-1-yl)methyl]-4*H*-1,2,4-triazole-3-thiol and 4-(4-methoxyphenyl)-5-[(2-methyl-1*H*-benzimidazol-1-yl)methyl]-4*H*-1,2,4-triazole-3-thiol were synthesized and investigated as inhibitors for mild steel corrosion in 15% HCl solution. It was found that the inhibition efficiency of these inhibitors increased with increasing concentration. This showed that all three studied inhibitors were of mixed type in nature [27]

- The Corrosion inhibition by triazole derivative 1-(2-pyrrolecarbonyl) benzotriazole (PCBT) and phosphono derivative 2-Phosphonoacetic acid (2-PAA) on mild steel in ground water media has been investigated. The experimental results obtained reveal that among the two inhibitors PCBT is the best effective inhibitor than 2-PAA. The variation in inhibitive efficiency mainly depends on the type and nature of the substituent present in the inhibitor molecule. [28].

- The inhibition effect of   N-benzyl –N'-phenyl thiourea (BPTU) on the corrosion of mild steel in 0.01 and 0.05 N HCl medium was investigated. BPTU is an efficient anodic inhibitor with greater than 94% of efficiency in the range of temperature studied [29].

- The corrosion inhibition of mild steel in 1M HCl and 0.5M H_2SO_4 by L-Cysteine (Cys) have been studied by Using electrochemical impedance spectroscopy (EIS) measurements and molecular dynamics (MD) simulation. The results show that the inhibition efficiency increase with the increase of Cys concentration in both acids, and the higher inhibition efficiency is obtained in 1 M HCl. The adsorption of Cys molecules on the mild steel surface obeys Langmuir adsorption isotherm in both acids, and occurs spontaneously. The MD simulation results show that with the adsorption of chloride ions has the higher negative interaction energy comparing to the case of the adsorption of sulfate ions in the Fe + anion + Cys simulated system. [30].

- It is also reported that smaller inhibitor molecule are efficient because they facilitates electronic interactions and impede steric effects. But many of them are toxic even though they exhibit good inhibition action [31].

- The attempts have been also done for drugs as inhibitors [32], Tryptamine [33], ketoconazole [34], sulfa drugs [35], antibacterial drugs [36], rhodanine azosulfa drugs [37] tacrine [38] have been reported to be good inhibitor. The planarity of the molecule, besides other factors, decides the efficiency of an inhibitor [39].

- The synergistic effects of some ions such as KI, KCl and KBr on the corrosion inhibition of zinc in 0.4 M HCl by tetrahydrocarbazole derivatives compounds were investigated. Results obtained show that these compounds are good inhibitors and their inhibition efficiencies (IE %) increase with the increase of inhibitor concentration [40].

- The effect of a (N-[(1E)-(4-methoxphenyl) methylene] hydrazinecarbothioamide, called ATSC, with chelating groups, on the corrosion behavior of zinc was investigated. Electrochemical study of the zinc specimens was carried out in aqueous electrolyte containing 0.2 M Na_2SO_4 and 0.2 M NaCl maintained at pH 5 [41].

- The corrosion inhibition characteristics of acetyl coumarine (AC), bromo acetyl coumarine (BAC) and thiazole derivatives (BTMQ and BTCQ) on the corrosion of zinc in 0·1 M HCl solution were investigated by weight loss, potentiodynamic polarization and impedance techniques [42].

- Some Heterocyclic compounds (namely oxadiazole, thiadiazole,and triazole derivatives) show the as corrosion inhibitors for mild steel in hydrochloric acid medium [43]. The inhibition effect of heterocyclic organic derivative viz., 2-(Benzothiazol-2-ylsulfanyl)-1-(5- methyl-thiazol-2-ylamino)-ethanol was evaluated against mild steel corrosion in 1 N HCl solution [44].

- The corrosive behavior of zinc in HCl solution containing various concentrations of glutaraldehyde GTD, glycine GLN, methionine MTN, and their condensation products formed between GTD + GLN (CP1) and GTD + MTN (CP2) was investigate[45].

- Corrosion inhibition of mild steel by nevirapine, an antiretroviral has been investigated using potentiodynamic polarization, electrochemical impedance spectroscopy technique and weight loss methods [46].

- Inhibitor action of Risperidone, a drug, on mild steel in 0.5N HCl and 0.5N H2SO4 solutions has been investigated at different temperatures 298K, 308K and 318K. The inhibition efficiencies have been evaluated at different concentrations of the inhibitor [47].

- Benzotriazole derivatives, namely, N-[1-(benzotriazolo-1-yl)alkyl] aryl amine (BTMA), N-[1-(benzotriazolo-1-yl)aryl] aryl amine (BTBA), and 1-hydroxy methyl benzotriazole (HBTA), were synthesized and their inhibition behaviour on mild steel in 0.5 M H_2SO_4 at room temperature was investigated by various techniques[48]. The corrosion behavior of mild steel samples immersed in 1, 0.1, 0.01 and 0.001% Na_2SO_4 aqueous solutions is evaluated [49].

- The electroplating of zinc is carried out in the presence of 3, 4, 5-Trimethoxy benzaldehyde from a chloride bath. Corrosion resistance test indicated good protection of steel by the coating [50].

- Polyaniline (PANI) thin films were electrochemically deposited by cyclic voltammetry or stainless steel electrode previously covered by a thin film of polyvinyl acetate (PVAc). The corrosion resistance of PANI covered stainless steel substrates was estimated by using potentiodynamic polarization curves and its linear polarization resistance (LPR) was measured in 0.5 M H2SO4, 0.5 M NaCl and 0.5 M NaOH aqueous solutions at room temperature. The results indicate that the PANI-PVAc films did improve the corrosion resistance of the stainless steel in NaOH, behaving even worst, in the case of PANI film, than the uncoated substrate. In H_2SO_4 both PANI and PANI-PVAc coatings gave good protection for the stainless steel electrode, with a slightly better performance of PANI-PVAc than PANI. In NaCl solution both PANI and PANI-PVAc films provided a good protection against corrosion. The better performance of PANI-PVAc coatings for corrosion protection in basic media may be due to its major chemical stability compared to simple PANI films, which lose their conductivity in high pH solutions. The Ecorr (free corrosion potential) value of the coated substrate was in the passive region of the uncoated substrate in acidic environment but in the active region in neutral or basic environment [51]

- The corrosion and inhibitor adsorption processes in mild steel/1-methyl-4[4(-X)-styryl] pyridinium iodides (X: H, CH3 and OCH3)/hydrochloric acid systems was studied at different temperatures (25–60 ◦C). It was found that the studied compounds exhibit a very good performance as inhibitors for mild steel corrosion in 1.5M HCl. Results show that the inhibition efficiency increases with decreasing temperature and increasing concentration of inhibitors [52].

- Corrosion inhibition by triazole derivative 1-(2-pyrrolecarbonyl)benzotriazole (PCBT) and phosphono derivative 2-Phosphonoacetic acid (2-PAA) on mild steel in ground water media has been investigated by weight loss method, potentio dynamic polarization and electrochemical impedance spectroscopy (EIS). The experimental results obtained reveal that among the two inhibitors PCBT is the best effective inhibitor than 2-PAA. The variation in inhibitive efficiency mainly depends on the type and nature of the substituents present in the inhibitor molecule. Inhibition efficiency (IE %) values obtained from various methods used are in good agreement [53].

- The corrosion inhibition of mild steel in 1M HCl solution by a synthesized compound (3-benzoylmethyl benzimidazolium hexafluoroantimonate) was investigated electrochemically and by weight loss experiments. The percentage inhibition increased with the increase of the concentration of the inhibitor and reached about 98% .The percentage inhibition decreased with the increase of temperature. This compound was found to be a very good corrosion inhibitor due to the presence of nitrogen in benzimidazole and phenyl ring [54].

- All the investigated inhibitors have some draw backs and some of them are listed below

  a) They are toxic in nature

  b) The corrosion inhibition efficiencies are decreased with temperature

  c) They need more concentration

  d) Corrosion Mechanism have been not developed

  e) Theoretical studies have not been done

- In our proposed work attention will be taken and try to give the solution for the above draw backs

References

- [1] L.L. Shreir, R.A.Jarman, G.T.Buratein,Corr. 1 (1994) 1- 6.

- [2] R.A.Prabhu, A.V.Shanbhag ,T.V.Venkatesha, J.App.Electrochem. 37 (2007) 491-497.

- [3] S.K.Rajappa, T.V.Venkatesha, Ind.J.Eng,Mat. Sci. 9 (2002) 213-217.

- [4] Shan-ZhuoYao, XiaoHuiiJiang, Li-MeiZhou,Ya-JuanLv,Xing-QiHu. 104 (2007) 301-305.

- [5] S.Tamil Selvi, V.Raman, N.Rajendran , J. App. Electrochem.. 33 (2003) 1175- 1182.

- [6] Mars G. Fontana, mcgraw-hill Inter. Edi.33 (1987) 126-133.

- [7] R.A.Prabhu,T.V.venkatesh, A.V.Shanubhag, B.M.Praveen, G.M.Kulkarni, R.G.Kalkhambar, Mat.chem.phys.108 (2008) 283-289.

- [8] S.M.A.Hosseini, A.Azimi, corr.sci.51 (2009) 728-732.

- [9] S.Kertit, J.Aride, A.Ben-Bachir, A.Sghiri, A.Elkoly.M.Etman, J.App.Electrochem.19 (1989) 83-89.

- [10] Lin Wang, Corr.Sci. 43 (2001)1637-1644.

- [11] K.C.Emeregul, O Atakol, Mat. Chem.Phys.82 (2003) 188-193.

- [12] S. Bilgie, N. Caliskan, App. Surf. Sci.152 (1999) 107-112.

- [13] M.G.Hosseini, S.F.L.Mertens, M.Gorbani,M.R.Arshadi, Mat.Chem.Phys.78(2003) 800.-805.

- [14] H. K.Kadiya , R. T. Vashi, Hind. Pub.Corp. 6 (2009) 1240-1246.

- [15] H. K. Kadiya , R. T. Vashi, Hind. Pub. Corp.6 (2009) 480-484.

- [16] M.N. Desai, M.B. Desai, C. B. Shah, S .M. Desai, Corr.Sci. 26 (1986) 827-831.

- [17] M.Sahin, S.Bilgic,H.Yilmaz, App. Surf. Sci. 195 (2002) 1-7.

- [18] Y.K.Agrawal,J.D.Talati,M.D.Shah,M.N.Desai ,N.K.Shah,Corr.Sci.46 (2004) 633-651.

- [19] H.M.Bhajiwala , R.T.Vashi, B.Electrochem.18 (2002) 261-266.

- [20] I.A.Abdel Wahab,O.R.Khalifa,Mostafa , B.AboEl Khair,A. J.Chem.5 (1993) 1084-1097.

- [21] A.A. Aksut, A.N. Onal, B. of Electrochem.11 (1995) 513-521.

- [22] I.Z. Selim, B. Electrochem.13 (1997) 385-391.

- [23] E. Khamis, Corr.46 (1990) 476-482.

- [24] 1. S.Manov, F.Noli, A.M. Lamazouere, L. Aries, J. App. Electrochem., 29 (1999) 995-1000.

- [25] J.N. Gaur and B.L. Jain, J. Electrochem. Soc. Ind. 27 (1978) 117-122.

- [26] A.G. Gad Allah, M.M. Hefny, S.A. Salih,M.S. El-Basiouny, Corr. 45 (1989) 574-580.

- [27] K.F. Khaled,App. surf.sci. 252 (2006) 4120-4128.

- [28] MahendraYadav, Debasis Behra, Sumithkumar, RajeshRajansinha, Ind.Eng.Chem.52 (2013) 6318-6328.

- [29] V.Manivannan, N.Chithralekha, Cor.J. Sci.1 (2012) 17-19.

- [30] S.Divakar shetty, Prakash shetty, Ind.J.Chem.tech.15 (2008) 216-220.

- [31] Hui Cang,Zhenghao Fei,Wenyan Shil,Oi Xul,Int.J.Electrochem.Sci.7 (2012) 10121-10131.

- [32] S.A. Umoren, U.F. Ekanem, Chem. Eng. Comm. 197 (2010) 1339–1356.

- [33] Ashish Kumar Singh, M.A. Quraishi, Corr. Sci. 52(2010)152–160.

- [34] Pongsak.Lowmunkhong, Dusit.Ungthararak, Pakawadee Sutthivaiyakit, Corr.Sci. 52 (2010) 30–36.

- [35] I.B. Obot, N.O. Obi-Egbedi, Corr. Sci. 52 (2010) 198–204.

- [36] M.M. El-Naggar, Corr. Sci. 49 (2007) 2226–2236.

- [37] M. Abdallah, Corr. Sci. 46 (2004) 1981–1996.

- [38] M. Abdallah, Corr. Sci. 44 (2002) 717–728.

- [39] S.E. Nataraja, T.V. Venkatesha , H.C. Tandon ,Corr. Sci. 60 (2012) 214–223.

- [40] M. Abdallah, S.T.Atwa, M.M.Salemand A.S. Fouda Int. J. Electrochem. Sci., 8(2013) 10001 – 10021.

- [41] R.A.Prabhu,T.V.Venkatesh,B.M.Praveen, ISRN Metallurgy 2012 (2012) 1-7 .

- [42] A V Shanbhag, T V Venkatesha, R A Prabhu, B M Praveen, B. Mat.Sci.34 (2011) 571-576.

- [43] F. Bentiss, M. Lagrenée, J. Mat. Environ. Sci. 2 (2011) 13-17.

- [44] Bhupendra M Mistry,Niketan S Patel,Smita Jauhari, Arc.App.Sci.Res.3 (2011) 300-308.

- [45] Shanthamma Kampalappa Rajappa, Thimmappa V. Venkatesha, Turk.J.Chem. 27 (2003) 189 -196.

- [46] Ishwara Bhat, Vijaya D. P. Alva, J. Kor.Chem.Soc. 55 (2011) 836-841.

- [47] R.A. Prabhu, A.V. Shanbhagb T.V. Venkatesha,B. Electrochem.22 (2006) 225–233.

- [48] Tamil selvi, V.Rama,N. Rajendran,J. App. Electrochem. 33 (2003) 1175–1182.

- [49] S. Arzola1, M.E. Palomar-Pardave, J. Genesca, J.App. Electrochem.33 (2003)1223–1231.

- [50] Yanjerappa Arthoba Naik, Thimmappa Venkatarangaiah Venkatesha,Perdur Vasudeva   Nayak,Turk.J.Chem.26 (2002) 725 -733.

- [51] M.E. Nicho, hailin hu, J.G. Gonza ,Lez-rodriguez,V.M,Salinas-Bravo, J.App.Electrochem. 36 (2006) 153–160.

- [52] Ehteram A. Noor, Aisha H. Al-Moubaraki,Mat.Chem. Phy. 110 (2008) 145–154.

- [53] V. Manivannan,N. Chithralekha ,Corr. J. Sci. 1 (2012) 17-19.

- [54] Ayssar Nahl´e, Ideisan I.Abu-Abdoun, Ibrahim Abdel-Rahman,Int.J. Corr. 2012 (2012) 1-10

Objectives of the Work

- Metals and alloys are increasingly used in newer fields besides their conventional applications. Every industry makes use of one or more metals and alloys in its functioning. The unending demand for the structural materials never appears to die down as activities have crossed horizons. Wide variety of material and alloy combination with innumerable compositional make up has made their way in to market. Therefore, it is rightly said that metals are the back bone of industrial progress. As metals are universal in their use, its corrosion is natural, fundamental and preventable. The study of corrosion of metals and its control occupies a prominent place in academic and industrial sector and in the field of material science.

- The development of new technologies employs the use of some uncommon and expensive metals. As newer materials are entering the arena, newer aggressive pollutants are joining the atmosphere, in a proportions which was never been imagined. These new pollutants join hands with residual pollutants to augment corrosion. Thus corrosion of metals received a prime attention of material scientists and chemists as corrosion results in a tremendous financial loss both directly and indirectly. Corrosion of metals and its prevention is an international problem and every country encourages research and investigation on the subject to develop better remedial measures for combating corrosion.

- It is well documented in the literature and still practiced is that, it can be controlled by the use of inhibitors. Various types of chemical inhibitors like organic and inorganic compounds, pharmaceuticals, drugs, dyes and even plant extracts are developed and continue to developed better in the coming days to come. The science behind the use of corrosion inhibitors has been investigated in details and well-established which claim that it is rather a simplest much cheaper technique. The inhibitor is added to the environment in situ.

- In spite of this, their toxicity and environmental damage is the issue of great concern. Hence, hitherto widely used inhibitors such as chromates, cyanides, etc., are now off the use. This leads a large scope for developing novel compounds, drugs and similar compounds with better characteristics. Still important is the academic issues behind them. The role and the mechanism of an inhibitor can only be understood with sound knowledge about it. For example, issues like the ambiguity about the size of the inhibitor molecule whether big or small molecules are efficient, can only be resolved through a comprehensive study. In this context, theoretical as well as experimental studies to understand the real science behind the inhibition action are the prime area of research in these days. Quantum chemical studies and molecular dynamics are the key areas in theoretical investigation. The advent of high configuration computers has revolutionized this branch of science. Huge numbers of softwares are available commercially in the market at very affordable price. Furthermore, very versatile and accurate methods are on the anvil which mimics the almost real working system. This encouraged scientists to take up research in these lines. Huge number of publications in peer reviewed journals shows the importance. The theoretical knowledge combined with experimental proof puts a researcher to the climax of understanding. Experimentally, various chemical and electrochemical methods have been used for a long time to study the corrosion inhibition ability of the compounds. Corrosion rate, inhibition efficiency, inhibition mechanism, adsorption profile, etc. could be evaluated by mass loss, Tafel extrapolation, and linear polarization and Electrochemical Impedance spectroscopic methods. Surface characteristics could be investigated by using Scanning Electron Microscopy (SEM).

- Clear consensus is there among the scientists to evaluate science behind corrosion inhibitors by both theoretical and experimental studies. A sound knowledge can be experienced by comparing, correlating and come out with reasoning between them.

- Main objectives of the work has been given below

    a) To scan large number of electro active compounds for corrosion inhibition action

    b) Selection of electro active species is made such that they bear maximum efficiency

    c) Classify and group them on the basis of structure, functional group, solubility, Inhibition efficiency, mode of adsorption, etc

    d) Comparing the inhibition action of these compounds (group wise and as a whole)

    e) Establish the Mechanism of the Inhibition action

    f) Quantum and thermodynamic studies for the inhibitors

    g) Investigating the surface morphology of corroded specimens by spectroscopic technique

    h) Find the best inhibitor among the compounds that is investigated

    i) Optimization of concentration of the inhibitor which exhibit maximum inhibition efficiency

Methodology

- Preparation of Steel or zinc specimens

- Preparation of Corrosive/acid medium

- Selection of suitable inhibitors

- Measurement of corrosion rate by weight loss, Tafel plot, linear polarization, impedance method

- Calculation of the thermodynamic parameters

- Investigation of corrosion rate by quantum studies

- To study surface characteristics by Scanning electron microscope spectral studies

- Assigning a mechanism for the inhibition for inhibitor

YOUR KNOWLEDGE HAS VALUE

- We will publish your bachelor's and
  master's thesis, essays and papers

- Your own eBook and book -
  sold worldwide in all relevant shops

- Earn money with each sale

Upload your text at www.GRIN.com
and publish for free